Laura Fagan

A study on the relationship between Video Gaming and Language Learning

Laura Fagan

A study on the relationship between Video Gaming and Language Learning

LAP LAMBERT Academic Publishing

Impressum / Imprint

Bibliografische Information der Deutschen Nationalbibliothek: Die Deutsche Nationalbibliothek verzeichnet diese Publikation in der Deutschen Nationalbibliografie; detaillierte bibliografische Daten sind im Internet über http://dnb.d-nb.de abrufbar.

Alle in diesem Buch genannten Marken und Produktnamen unterliegen warenzeichen-, marken- oder patentrechtlichem Schutz bzw. sind Warenzeichen oder eingetragene Warenzeichen der jeweiligen Inhaber. Die Wiedergabe von Marken, Produktnamen, Gebrauchsnamen, Handelsnamen, Warenbezeichnungen u.s.w. in diesem Werk berechtigt auch ohne besondere Kennzeichnung nicht zu der Annahme, dass solche Namen im Sinne der Warenzeichen- und Markenschutzgesetzgebung als frei zu betrachten wären und daher von jedermann benutzt werden dürften.

Bibliographic information published by the Deutsche Nationalbibliothek: The Deutsche Nationalbibliothek lists this publication in the Deutsche Nationalbibliografie; detailed bibliographic data are available in the Internet at http://dnb.d-nb.de.

Any brand names and product names mentioned in this book are subject to trademark, brand or patent protection and are trademarks or registered trademarks of their respective holders. The use of brand names, product names, common names, trade names, product descriptions etc. even without a particular marking in this work is in no way to be construed to mean that such names may be regarded as unrestricted in respect of trademark and brand protection legislation and could thus be used by anyone.

Coverbild / Cover image: www.ingimage.com

Verlag / Publisher:
LAP LAMBERT Academic Publishing
ist ein Imprint der / is a trademark of
OmniScriptum GmbH & Co. KG
Heinrich-Böcking-Str. 6-8, 66121 Saarbrücken, Deutschland / Germany
Email: info@lap-publishing.com

Herstellung: siehe letzte Seite /
Printed at: see last page
ISBN: 978-3-659-69504-9

Zugl. / Approved by: Dublin, Dublin Business School, 2014

Copyright © 2015 OmniScriptum GmbH & Co. KG
Alle Rechte vorbehalten. / All rights reserved. Saarbrücken 2015

Contents

Acknowledgements 2
Introduction 4
   Theory 5
   Case Studies in relation to language learning and video gaming 9
   Theoretical Conclusion 11
Hypothesis 13
Method 14
   Design 14
   Quantitative Data 21
   Data about the Test Scores 28
Correlations 32
Hypothesis 1 32
   Hypothesis 2 35
   Hypothesis 3 38
   Hypothesis 4 42
   Hypothesis 5 45
   General Video Game Players 49
   General Streamers 53
   Professional Streamers 55
Discussion 57
Further Research 63
Bibliography 64
Appendix 1 66
Appendix 2 67
Appendix 3 69

Acknowledgements

This thesis would not have been possible to complete without the help and assistance of numerous people. It would be impossible to name every single person who has helped me in this process but there are a few people who I would like to give a mention to:

I wish to acknowledge Anna Wolniak, my supervisor, for her unreserved patience, assistance and knowledge throughout this whole process.

I thank Paul Hollywood, my course coordinator for having faith in me and giving me the opportunity to do this project.

To all the library staff in Dublin Business School Ireland, particularly Maria Rogers; for their guidance, knowledge and patience for all the questions I asked regardless of what time of day it was.

I'd like to thank the staff of Cambridge English Language Assessment for permission to use one of their tests in order to establish a baseline for the participants of the survey. Without this critical piece, it would have been impossible for this piece of work to have the weight behind it that I wanted. Cambridge University has my unreserved thanks.

Finally, to my parents, my partner and my friends both near and far who have had to put up with me being buried in a book or a laptop for the last year. Without your endless love, support and understanding, and help in multiple aspects of this project would not have happened at all.

Abstract

In traditional gaming language, learning is not something that is commonly thought of in the video game industry or could arguably be considered in the design of traditional games. This dissertation is to look at traditional games to observe if there is an effect on language learning. What I intend to contribute to this body of research is to answer the question: "Is playing video games linked to the player's baseline English language level? My results, obtained via a combination of quantitative and qualitative research that addressed language levels, gaming habits, and personal experiences involving language learning and gaming, indicates inconclusive results due to conflicts between the quantitative data. I aim to explain the potential reasons behind this conflict. The implications of my results point to an avenue of focus for future research in several aspects of this issue. The immediate research potential of this is the address of the conflicts that have arisen and developing new measures. However, the list of future research from this thesis is inexhaustible.

Introduction

As we move further into the 21st century, video games are growing as a prominent technology. Video games are becoming more commonplace and more impactful in recent years – especially with consideration to their potential benefits, which will be discussed later in this dissertation. Evidence regarding the extent of video gaming's impact on society is reflected in the reports given by the Entertainment Software Association which include sales, demographics, and usage data. For example, the 2013 report reveals that 58% of Americans play videogames as compared to 53% in 2004 (Entertainment Association, 2004 and 2013). The demographics of 21st century gamers are also changing. As of 2013, the average age of players is 30 and 55% of American players are male and 45% are female (Entertainment Association, 2013). In comparison, the Association's 2004 report stated that the gender break down was 59% male and 39% female (Entertainment Association, 2013).

It's important to notice that in the Entertainment Association 2013 report that 52% of gamers say that they play games with others, either in person or online. This is a key element to consider in regards to language learning. This figure has increased since the Association's 2004 report which stated that 43% of players play online (Entertainment Association, 2013). In an overwhelming statistic, it shows that 77% of the gamers, who said that they played with others, said they did so for at least one hour a week (Entertainment Association, 2013).

Theory

When you exclude games specifically written for education purposes, language learning is not something that would be commonly thought of in the video game industry or could arguably be considered in the design of traditional games.

One of the aims of this dissertation is to look at traditional video games to observe if there is an effect on language learning, the possibility that non native English Speakers can learn English through video gaming. It could be argued that learning a language depends on your natural ability to learn a language and if you are already good at learning languages you might not benefit as much as those who struggle. However, it is felt amongst theorists like Gee that immersion is the key in order to facilitate learning, because language learning is a primal instinct in human beings, as communication is key to human survival. There is a collection of data to support this theory and there are possible theories that could be used to explain why this occurs. The argument that this dissertation is going to look at is whether language learning takes place in video games or not.

Pacer (2009) describes learning as "process by which experience produces a relatively enduring and adaptive change in an organism's capacity for behaviour." This definition describes an important difference between learning and performance as well as change in the capacity of behaviour. Pacer (2009) goes on to discuss the element of environment and its affect on learning. Pacer considers that the environment in which we learn influences adaptation in two significant ways; species adaptation and personal adaptation. Species adaptation concerns itself with natural selection, a process that is a guide to evolution. Personal adaptation occurs through learning.

Pacer (2009) then introduces the concepts of Habituation and Sensitization as part of an element of learning processes. Habituation and sensitization are important elements in learning as they define the increases and decreases in the strength of a response to a stimulus. This is a simple definition of the processes that are occurring – it allows the organism to respond appropriately in situations where repeated stimulus is important or threatening (this is sensitization), or when other stimuli are more important (habituation).

According to Thorndike, (as cited in Pacer, 2009) there are many ways in which we can be conditioned to learn. One particular law that could be considered in play is Thorndike's law of effect. The Law of Effect states that "responses followed by satisfying consequences will be strengthened, whereas those followed by annoying consequences will be weakened" (Pacer, 2009). Reinforcement comes when the responses are strengthened by the outcome that happens after them. The same thing applies for positive and negative stimuli – they will evoke some response which can then theoretically become a learned response. Operant extinction is the weakening and eventual disappearance of a response because it is no longer enforced. This could prove to be a key explanation in relation to the constructs of language learning.

It could also be argued that one of the primary principals behind language learning through video games, is observational learning. Observational learning is defined by Pacer (2009) as learning that occurs by observing behaviour of a model, a structure, or a system in which it is possible to learn.

Another element of language learning could be explained by Social-Cognitive Theory which was developed by Bandura (1964; 2004). The theory describes a four step process: attention, retention, reproduction and motivation. Bandura (1964; 2004) stated that observing successful models can increase self-efficacy and thus motivate people into performing the modelled behaviour. Modelling is often described as a key technique in learning for everyday situations.

According to Gee (2005), immersion is key in order to facilitate learning. The Merriam-Webster Dictionary defines immersion as "a method of learning a foreign language by being taught entirely in that language" or "complete involvement in some activity or interest." This focus on interest may appear to be an obvious statement, but it is a key point that is commonly overlooked in regards to language learning or any learning. It is a very simple principle: you have to be interested in what you do in order to learn.

There is a strong theoretical backing in relation to the immersion theory and also other theories in relation to second language learning. According to Mitchell et al (2013), it all begins with the Input Hypothesis (Krashen, 1982, 1985) and the earliest version of the Interaction Hypothesis (Long, 1981, 1983a, 1983b). Krashen's key idea is that "comprehensible input" is necessary and sufficient for language learning to take place. The Input Hypothesis proposes that if input is understood and if there is a sufficient amount of it, then learning is possible. However, this would be considered to be just looking at one side of any given interaction and that it pays very little attention to the workings of the device (pp.184-185). Interaction Theory (1996) took a more cognitive direction in relation to learning theory and was geared more towards learner attention and second language processing capacity as mediating factors. It is still unknown how these things are interoperated and processed. Swain's Output Hypothesis (pg. 187) is considered by Mitchell et al to be more balanced, as attention is paid to the input being received, to negotiations of both form and meaning, and to the productions of the learner.

Mitchel et al. (pg. 246) summarizes the complexities of this process by saying that major puzzles remain in relation to interaction theory due to a lack of transition from approach to a theory in explaining underlying cause-effects of the interaction approach. Additionally, Mitchel et al. discusses the importance of sociocultural theory which, in the 1990s, established itself as an active research programme within the field of second language learning, Mitchel et al. (pg. 249) considers that the reason for its popularity is because of the creative agenda for the renewal of second language practice, which has an appeal to educators due to its developmental opportunities that can be used in a classroom environment. One of the Sociocultural theory's major flaws is that it lacks empirical validity due to its nature, so measures need to be created around the sociocultural theory in order for it to become valid.

Comparatively, Gee's "Good Video Games and Good Learning" (2005) suggests that good video games incorporate good learning principles which are supported by current research in cognitive science (pg. 34). Furthermore, Gee proposes that no one would buy games if they could not be learned, and players would not accept easy, dumbed down, or short games. Gee's findings show that challenge and learning are a large part of what makes a good video game motivating and entertaining. He then goes on to identify learning principles that good games incorporate.

Another contributing factor to language learning could be seen in Gees' description of the process of Identity. This is, according to Gee, the manner whereby good games capture players through Identity. Gee states that no deep learning could take place unless learners make a commitment to themselves. Players become committed to the new virtual world in which they live and learn through their new identity (pg. 34). Furthermore, Gee introduces Interaction with a notion, that Plato discussed in Phaedrus, which is that books are passive; you cannot get them to talk back to you in real time dialogue. Gee argues that games do talk back and that indeed nothing can happen without the player deciding it first. Words and deeds are all taking place within the context of the interactive relationship between the player and the virtual world they are engaging with (pg. 34). Another important principal that Gee introduces is risk taking. It involves the concept that risk taking is essential in video gaming, and so is failure. Indeed, games encourage you to fail in order to learn. This is not an attitude that is encouraged in school (pg. 35). These elements could also be considered to be contributing factors in the process of language learning through video games.

These fundamental theories are the basis of Gee's paper of "Good Video Games and Good Learning" (2005). All give a sense of Agency or Ownership which is unheard of in a schooling environment. This means that the player can take charge of what they do in a game and how they do it. This is not an option in traditional learning scenario.

Case Studies in relation to language learning and video gaming

As the theoretical groundings for generalised learning and second language learning have been established, it is time to look upon some case studies in which video gaming has been used as an aid to learn a second language. This will provide a foundation from which to discuss the application of video games as a tool for language learning and to determine if it works as the theories suggested. It is important to note that the games mentioned in these case studies are games that were not designed for educational purposes.

One such study, Engaging 21st Century Adolescents: Video Games in the Reading Classroom (Adams, 2009) discusses video game use in a classroom environment and how its use affected the development of learners English. A teacher encourages two children with lower levels of English comprehension to play video games in order to increase their confidence with English text (pg. 57). Adams reports that her experiment is in line with the findings of Simpson and Clem (cited in Adams 2009) that "students learn more and more rapidly when actively engaged" (p. 55). Both students studied by Adams felt more confident with general reading material as a result of playing the video game "Never Winter Nights."

"US China E-Language Project: A Study of a Gaming Approach to English Language Learning for Middle School Students" (Green, Sha, and Liu, 2011) is, to date, the most sizeable study on the relationship between video gaming and English language learning (pg. ix). This study examines improvement in English language learning by testing the students at two periods at the start of the year and at the end using "The Forgotten World" video game. The project was broken down into a control group and a group who were playing "The Forgotten World." Prior to the study, all of the students were in Standard English classes. Students in the "Forgotten World" group who started off with lower statistical levels of English i.e. the students who previously were not doing well in their English language class. They increased their levels more than the control group who were not playing the game. There was no change in groups with higher baseline language proficiency levels. There was a higher reported motivation for learning English recorded in treated schools than in control schools, and 95% of teachers in treated schools had changed how they thought of teaching (pg. x, pg. 31). However, some teachers commented that using video games was unsuitable preparation for ninth grade assessments. Green, Sha, and Liu (2011) is significant because it was undertaken in actual classroom settings, and it provides a reasonably large sample of more than 2000 learners. As such, the findings add perspective regarding who will benefit the most and least from using video games in language learning.

Additionally, Rankin, Gold and Gooch (2006) detailed the effect of 3D Role-Playing Games as a language learning tool. Their study, 3D Role-Playing Games as a language learning tool, utilizes Ever Quest 2 as a learning tool for English as a second language. The results demonstrate that Ever Quest 2 players who were either in the intermediate or advanced level of English as a second language, increased their English vocabulary by 40% as a result of game play interactions with non-playing characters (NPCs). Intermediate and Advanced students of English as a second language practiced conversation with playing characters which caused a 100% increase in chat messages during the sessions.

A comparative study to keep in mind with the qualitative work done in this dissertation is the work completed by Tugurt and Irgin (2009). They took 10 students both in primary and secondary school in Mersin, Turkey. The data collected was through observations, semi-structured interviews and phenomenological data analysis steps. It was noted in the beginning of the report that many school children spend a lot of time in internet cafés gaming and interacting with their peers and other players, through video games. Considering the participants overwhelming response towards video gaming and learning English through the medium of video games, the implications of this report puts into light integrating computer games into English Language Teaching curriculum.

Theoretical Conclusion

Taking the above case studies into consideration there is a strong argument for video gaming to have an effect on language learning. The majority of the case studies I have referred to could be subjected to the Observational Learning theory and the Social Learning theory. However all of these components by themselves do not make a proven theory. What is it about these theories and these reported cases that would equate this to being a concept for future consideration? What are the limitations of each of the major studies that have been discussed?

Adams' (2009) paper reports in line with the findings of Simpson and Clem. However, it makes no distinction between virtual literacy and information literacy. Another flaw in the article is that the data of the reported improvement is not detailed, which raises concerns about the actual improvement of the child, or if increased confidence would equate to higher performance due to its sample size. This study's findings indicate that further research into this area is warranted to re-create similar patterns in relation to video gaming and increased English.

While being the largest study to date the main criticism of the Green et al. (2011) was its sample size. The report states that the sample size of approximately 2000 participants may be questioned due to technical difficulties and a lack of accessibility to computer which was reported (pg. 32). It is also important to note the cultural differences between Eastern and Western societies as the societal views on gaming are likely to differ which could affect the real life implications of the study.

Gee's (2005) study is entirely theoretical and there are no experiments or data in relation to the successes or failures of his theories. This applies to the majority of Gee's research that has been used in relation to the completion of this dissertation. While all of the theories in their essence make logical and rational sense as to what may happen, there is no proof that such things happen as they have not been experimentally tested.

To summarise what has been discussed so far: we have highlighted that there is a gaming community that has increased in size throughout the years and has the potential to keep growing. There are instances in which it has been recorded that exposure to these traditional games has resulted in improvement in language within certain conditions. However these lack theoretical backing as to why this occurs. There are theories that could provide an explanation as to why these things happen, but do not have any data or numerical backing to prove the concept.

What I intend to contribute to this research is to identify if there is a potential link between video gaming and language learning - "is playing video games linked to the player's baseline English language level?" I will provide my results via quantitative study in which participants were asked about their experiences with language and video gaming, and qualitative research in which participants were interviewed about their experiences with language and video gaming, in order to see if any distinct patterns can be seen between my data and the research that precedes me.

Hypothesis

On the basis of the above study of the existing literature there can be distinguishing features. There can be several assumptions which can be our hypothesis for this study.

1): There is no difference between time spent gaming and the participants Cambridge Score/self reported new words. (H_o)

With more time spent gaming the Cambridge Score will be higher/self reported new works score will be higher. (H_1)

2): There is no difference between the length of time that participants began gaming and participants Cambridge Score/self reported new Words. (H_o)

Length of time that Participants began gaming at would increase participants Cambridge score/ self reported new words. (H_1)

3): Gaming Platform does not affect language learning. (H_o)

Gaming Platform effects language learning. (H_1)

4): Type of game does not affect language learning. (H_o)

Type of game effects language learning. (H_1)

5): Playing with others does not affect language learning. (H_o)

Playing with others effects language learning. (H_1)

Method

Design

The research is divided into two parts, both qualitative and quantitative. With regards to the quantitative element of the research, it is important to get as large a sample as possible for the purpose of achieving the most accurate results. With a larger sample size to work from, it is possible to take various data points and compare them to uncover and analyse any potential patterns. For example, based on the research in this study, it would be possible to determine if the gamers' starting age or the number of hours spent gaming had any effect on their Cambridge score, in comparison to gamers of a different starting age and of time spent gaming. The qualitative study seeks to discover what individual gamers think of video gaming with regards to language learning and what personal impact gaming has had upon their lives.

The quantitative part of the study consists of a survey hosted on KwikSurvey (www.kwiksurvey.com). Participants obtained this survey through word-of-mouth and other social media resources. For example, participants who completed the survey often retweeted the survey via their Twitter accounts, post about the survey on Facebook, and/or link the survey on their Twitch account, which is the world's leading video platform and community for gamers, with more than 45 million visitors per month. (Twitch.tv, 2013). Overall, there were 884 people who completed the questionnaires.

The survey consists of 17 questions - all multiple-choice. Question 8 should be given particular note, as this research was conducted with the aid of Cambridge University. In accordance with Cambridge University policy, participants were informed that: "This is not a Cambridge English exam and the test scores and levels are very approximate. Your score on this test cannot be used as proof of a formal language qualification."

To answer Question 8, the participants were asked to go to the following URL: http://www.cambridgeenglish.org/test-your-english/adult

learners/index.aspx?page=survey&pagenum=1 and complete the English language test provided by Cambridge University. After completing the assessment, the participants were given their score range in a multiple checkbox. The below table (fig 1) from the Cambridge website gives the score ranges and suggests which test the participants could potentially take based on the results of the assessment.

Your score	Recommendation
6 to 10	Cambridge English: Key (KET)
11 to 12	Cambridge English: Key (KET) or Cambridge English: Preliminary (PET)
13 to 15	Cambridge English: Preliminary (PET)
16 to 17	Cambridge English: Preliminary (PET) or Cambridge English: First (FCE)
18 to 19	Cambridge English: First (FCE)
20 to 21	Cambridge English: First (FCE) or Cambridge English: Advanced (CAE)
22	Cambridge English: Advanced (CAE) or Cambridge English: Proficiency (CPE)
23 to 25	Cambridge English: Proficiency (CPE)

Fig.1 Table of Cambridge Score Range

The purpose of using this test in my research was to assess the participants' baseline linguistic ability. This is not meant to be used as a definitive figure, but as a guideline of the participants' potential linguistic ability. In relation to how the data is structured, the participants' English language level, as measured by the Cambridge test, forms the dependent variable and the independent variable(s) is gaming - time, type of games etc.

At the end of the quantitative survey, participants were asked if they would like to be involved in an interview for the qualitative survey and six people agreed to take part in the interview. They were asked a series of 17 questions in relation to their gaming experiences, the reasoning behind gaming and their opinions on using video gaming as an aid for learning a second language. The questions can be found in the Appendix 2. All questions were optional for the participants to answer. All due consideration was given to the participants and they were at all times made aware that they were under no obligation to answer any question if they preferred not to. For the purposes of this dissertation, the participants will be referred to as Alpha, Beta, Gamma, Delta, Epsilon, and Zeta.

Another set of interviews took place with three people who actively stream on the gaming broadcasting Twitch, to obtain their opinions on gaming and language learning. These people were asked to participate in the interview due to their experiences as seasoned streamers on Twitch. They were asked a series of 13 questions in relation to their gaming experiences, the reasoning behind gaming and their opinions on using video gaming as an aid for learning a second language. The questions can be found in the Appendix 3. All questions were optional for the participants to answer. All due consideration was given to the participants and they were at all times made aware that they were under no obligation to answer any question if they preferred not to. For the purposes of this dissertation, the participants will be referred to by their streaming names to differentiate themselves from the other interview set.

Another important element of the research design is that the interviews will be coded and then conceptualised, in order to produce themes and other connecting elements within the interviews. Interpretative phenomenological analysis will be used as it focuses on how a participant makes sense of their personal and social word, in this instance in relation to video games. (Smith & Osborn, 2003)

Participants

Participants were selected by means of convince sampling. The participants were solicited via Twitter and through contacts with Twitch. In turn they notified people about the survey and encouraged them to partake. The non-native English speakers and the Native English speakers were all obtained by this process. Altogether there were 884 participants; 620 Non Native English Speakers and 264 Native English Speakers. The participants' level of experience with video games is noted by the age they started playing video games. The majority of the participants stated that they started gaming between the ages of 0 years to 10 years old. For the purposes of this dissertation the Native English Speakers will be excluded as there is a presumed level of proficiency given that they are native speakers. The age range of all the participants was >17-60+. In total there were 827 males and 57 females (fig 6). In the Non Native English Speaker condition there were 581 Males and 39 females.

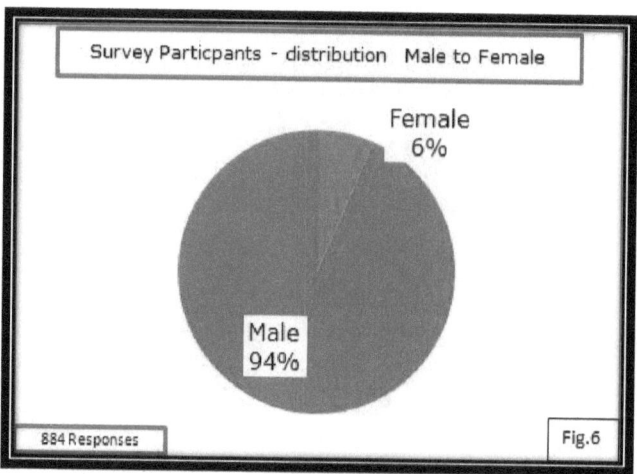

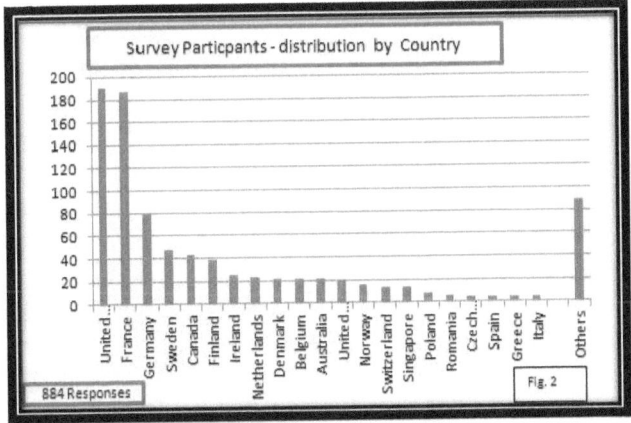

Fig. 2 details responses by country, with the greatest number of participants coming from the United States of America.

Fig. 3 shows how many participants responded to each question on the survey.

All participants in the quantitative study had to follow on-screen instructions as per the survey requirements. The maximum time required for the survey was five minutes.

Ethical Consideration

As far as ethical considerations for the quantitative study are concerned, the participants were informed that they were under no obligation to do the survey and that they were permitted to drop out at any time. Contact details were given to the participants at the start of the survey and at the end of the survey. If they have had any queries about the survey, they were able to withdraw at any stage. All possible care and consideration was given to the participants in relation to the data and the participants themselves.

Furthermore, equal ethical consideration was given to the participants of the qualitative study. The participants were all volunteers who contacted me of their own free will. All care and consideration was given towards the participants. Reassurance was given to them in respect to their privacy and whatever information they wish to impart via the interview. The following statement was made to the participants who wanted to take part in the interview process. The participants were given this statement before they took part in the interview to confirm their understanding that the information would be treated confidentially and would respect their privacy.

"For the qualitative part which you are helping me out with (you're free to drop out at any time, even after the interview has taken place) is your personal experience of gaming and how it has effected you and how you feel as a gamer that gaming has improved your language ability and if the community around the games have effected your interaction with the games and what your feelings are as to why this happens.

Just as long as you understand that whatever you say to me is completely confidential, you're not obliged to answer anything you don't want to answer, and you can tell me that you want stuff omitted from the report any time you want. Your name will not be disclosed nor any of your personal information (things like age range and gender are an exception for demographic reasons) and you will go under a false name for the purposes of the study. If you wish to know the name that you'll be under I'll be more than happy to tell you and give you a copy of my report upon submission."

For the quantitative study, a notice was provided at the beginning of the survey (fig 4) and the end of the survey (fig 5).

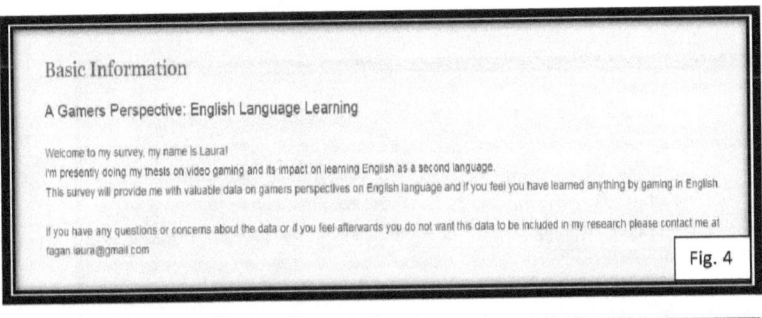

Fig. 4

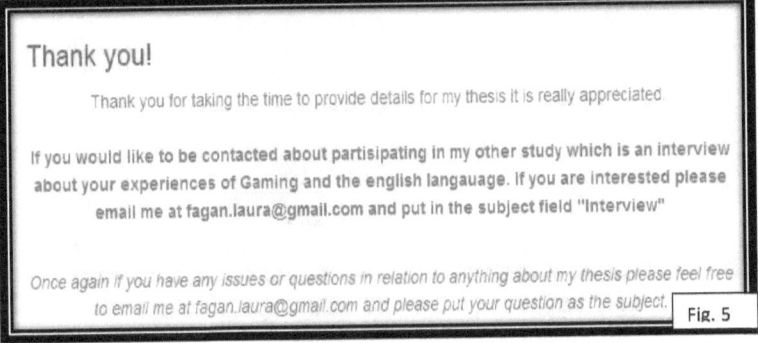

Fig. 5

Additionally, the participants of the interview were given this statement to read before (fig 4) they participated in the interview, as they had to consent to the interview: These statements were made to ensure the anonymity and privacy of the information provided by the participants. In the case of the streamer samples, all permissions were agreed to and the participants were under no obligation to reveal their streamer name unless they wished to do so.

RESULTS

Quantitative Data

In this section, the quantitative results from the study will be dealt with. It is broken down into introductory data which will introduce gender and age demographics as well as other establishing data.

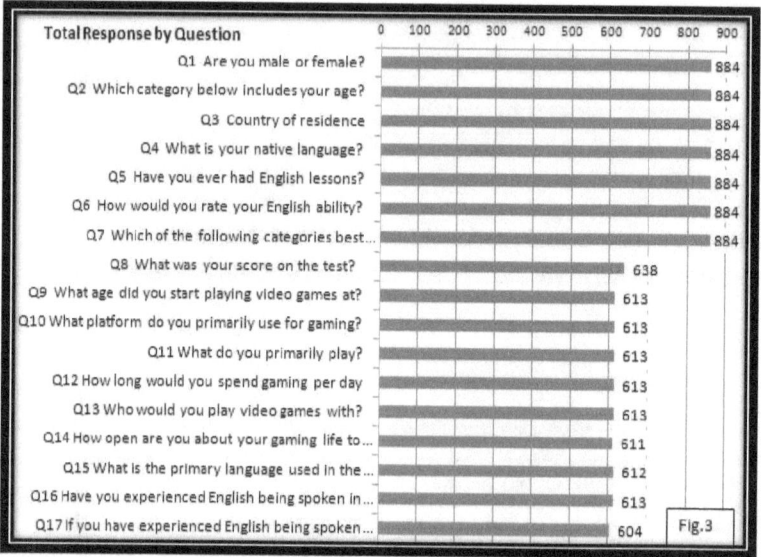

The above chart (Fig. 3) is a breakdown of all of the responses to each individual question. The maximum number of responses to each question was 884 while the minimum number of responses was 604. Question 8 is the Cambridge test results.

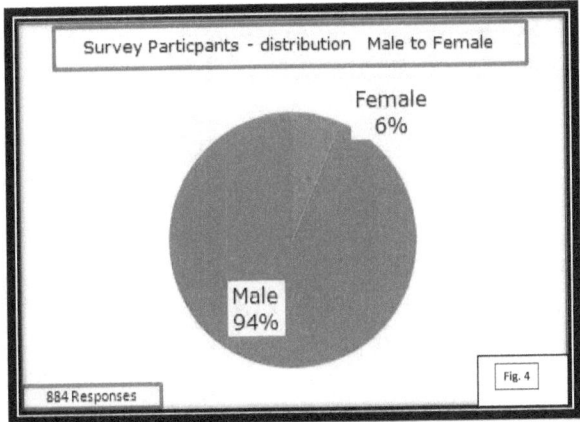

Out of 884 respondents, 94% of the respondents were male and 6% were female. (Fig. 4)

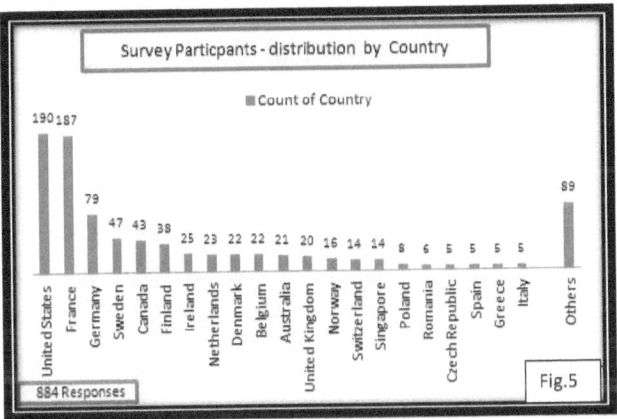

Fig. 5 shows the geographical distribution of the respondents. The top 21 countries were recorded as being the above graph. 89 participants, were distributed across 40 countries and these have been grouped in category "other" in Fig 5 above.

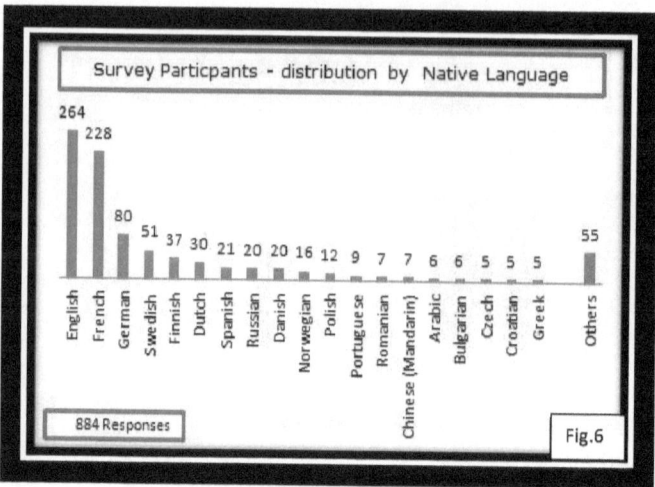

Fig.6

The above graph shows that most participants were native speakers of English. Discussion as to why this is the case will be dealt with in the Discussion section. The "Other" section was a collection of responses that was 4 and under. There were 55 other languages noted in this survey.

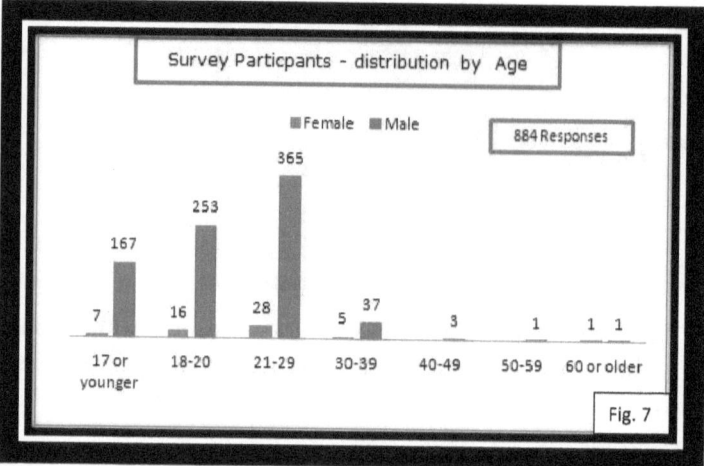

Fig. 7

The 21-29 age group bracket was dominant among the respondents, followed by the 18-20 age bracket. The age profile is not affected.

The majority of responses to the question regarding video gaming "what age did the participants start at" showed that overwhelmingly participants started playing video games around 0 – 10 years of age.

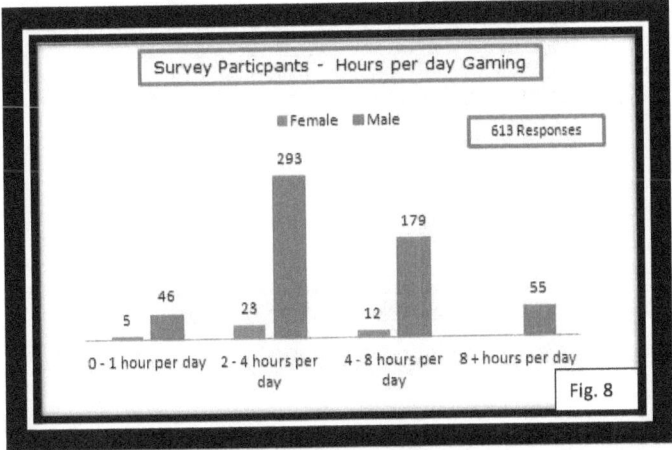
Fig. 8

51% male participants said that they spend 2 – 4 hours a day gaming while 58% female participants said they spend 2 – 4 hours a day gaming.

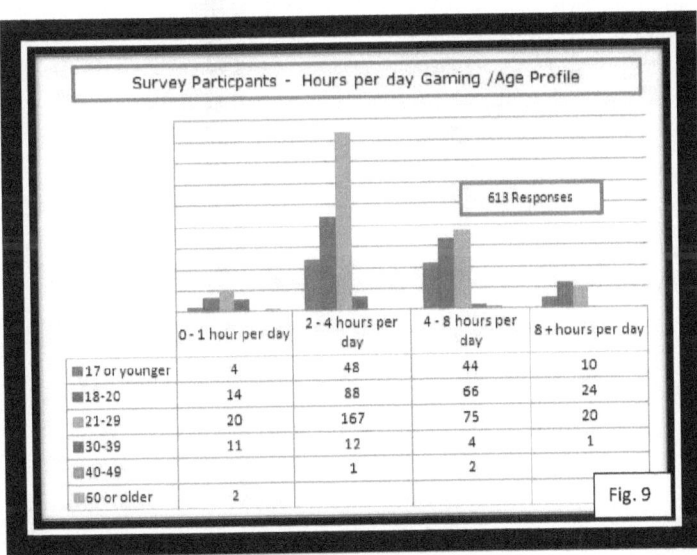
Fig. 9

The most time spent playing per day is the 2 – 4 hours reported by survey participants across all age profiles.

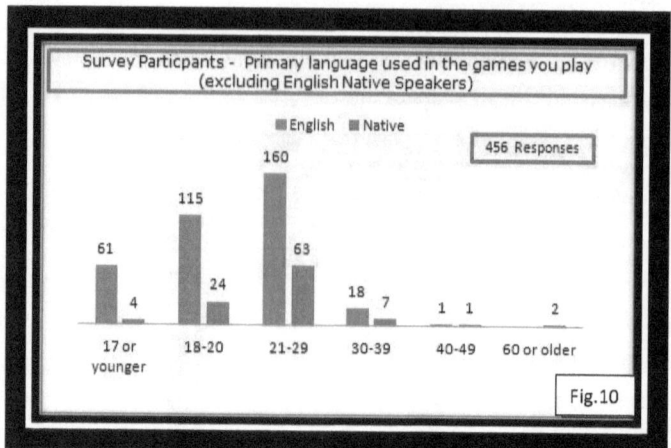

English language is noted in the 456 responses to be the most spoken language seen by the 17 to 30- 39 age groups.

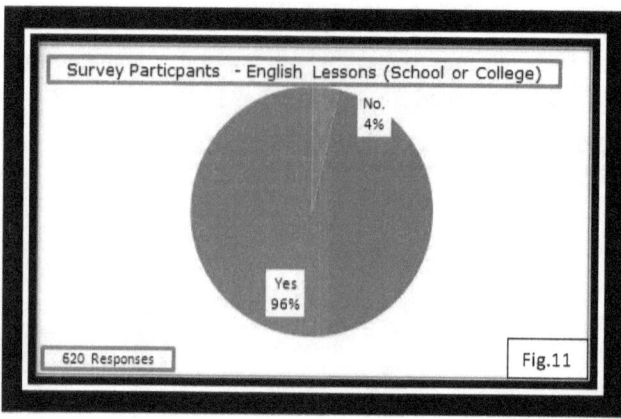

Out of 620 participants, over 96% report that they have had English lessons at some stage in their lives.

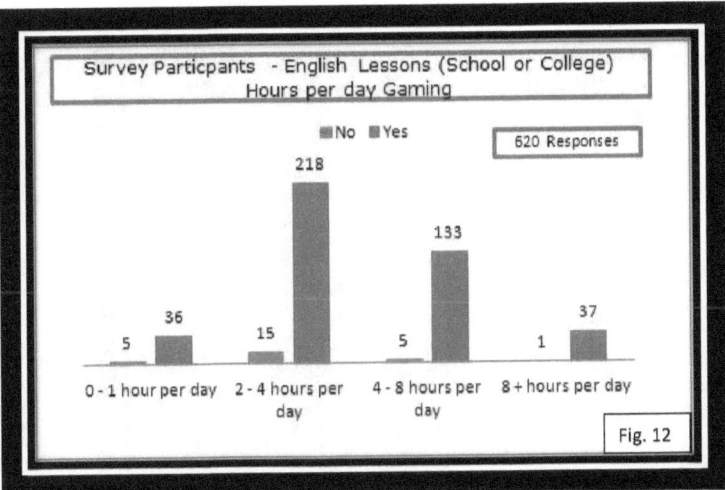

Fig. 12

Out of those who have had English lessons, the group who plays most daily are in the 2 – 4 hour range.

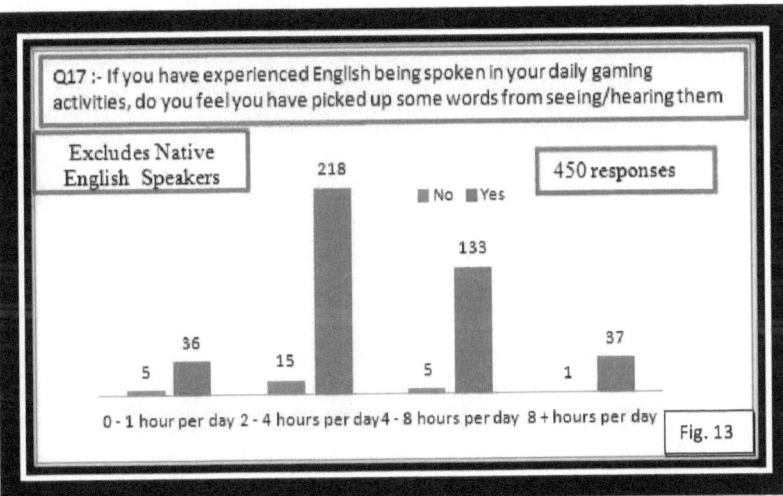

Fig. 13

Out of 450 responses the highest response of 218 say that they have felt they learned new words and this group played in the 2 – 4 hour range.

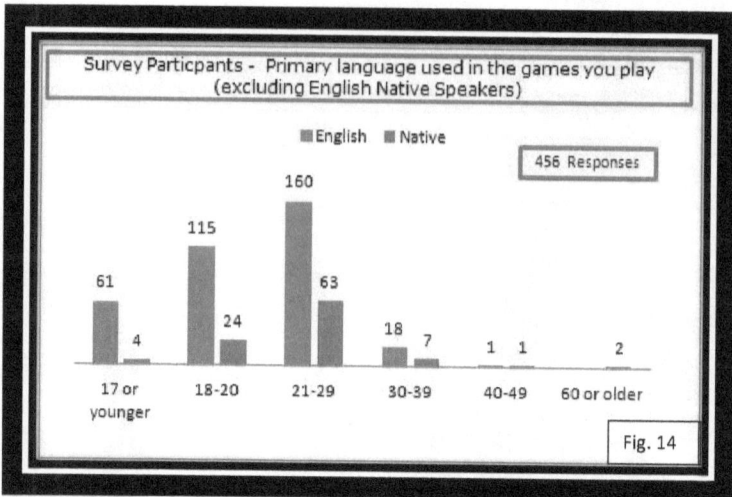

Fig. 14

The major language spoken by non-native English speakers in the 17 to 39 age group reported that English was the most predominant language.

Data about the Test Scores

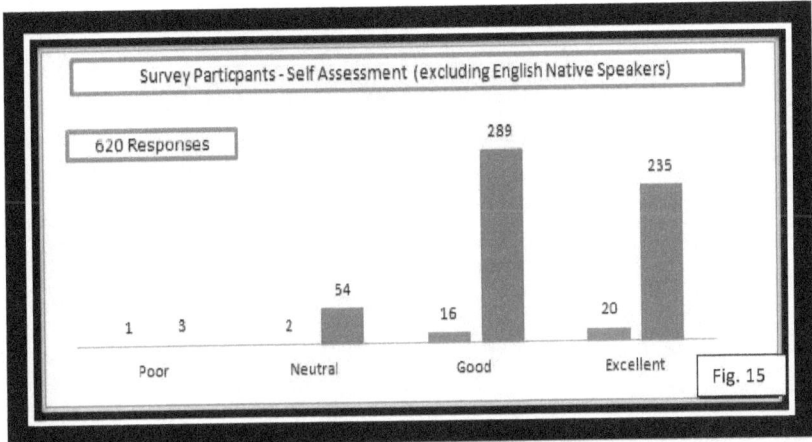

Fig. 15

Out of the self-reporting non-native English speakers, the majority of males were in the Good category. In women, they were in the Excellent category.

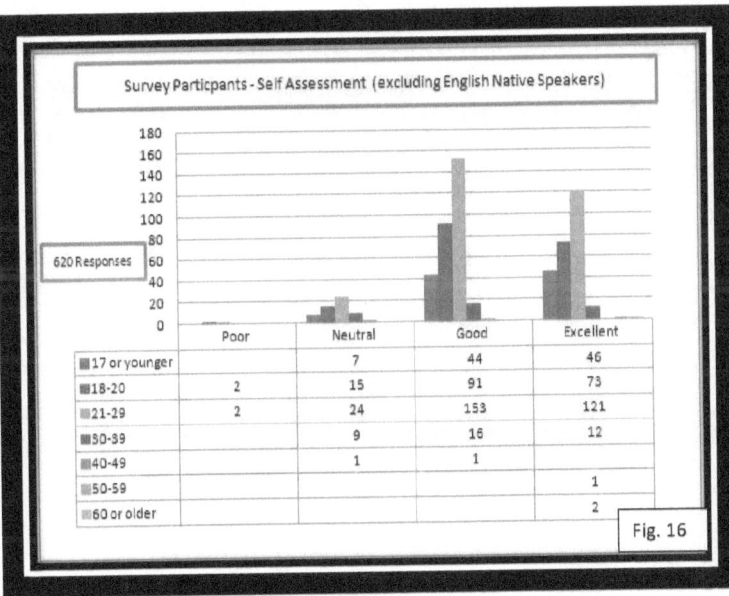

Fig. 16

On the whole, the most average self-assessment of non-native English speakers was in the Good category.

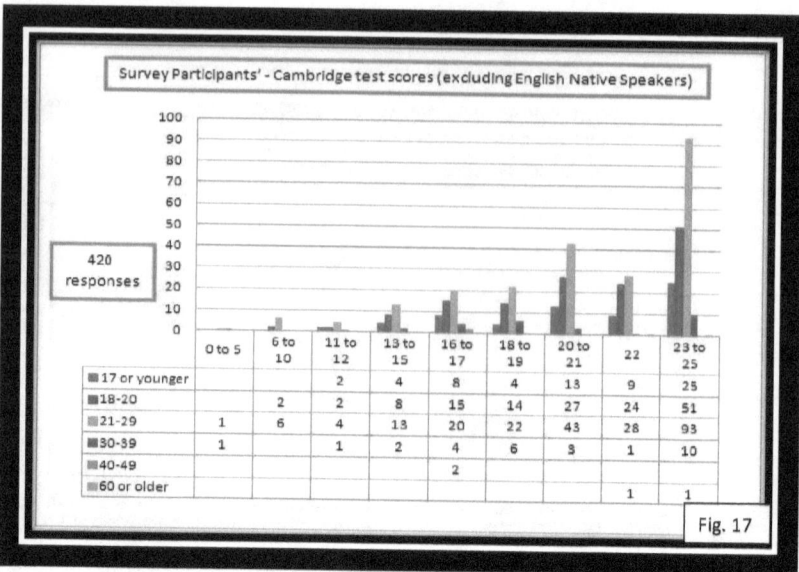

Fig. 17

Out of the 420 non-native speakers who responded to the test, the majority in all categories of age groups were in the 23- 25 range, which is the highest range on the Cambridge scale.

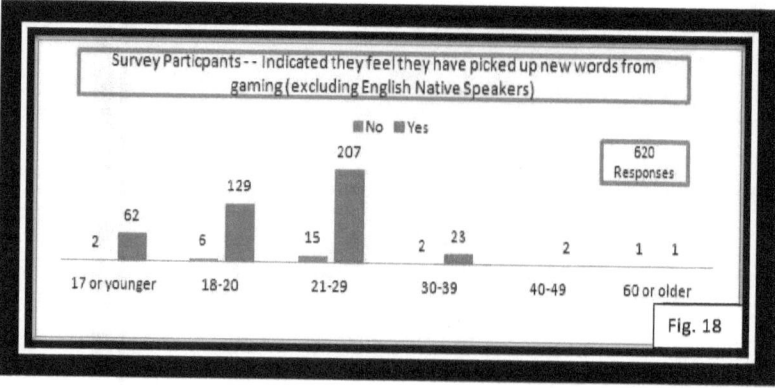

Fig. 18

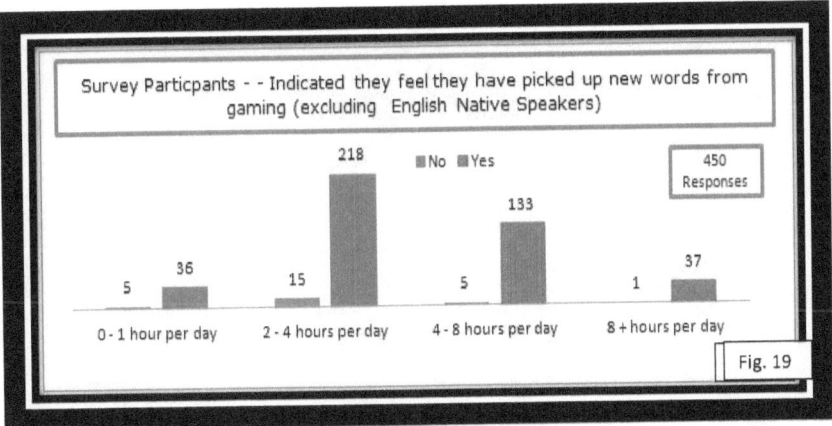

The majority of non-native English speakers say that they believed that they had picked up new words through gaming. The time the majority spent gaming was in the 2 – 4 hour range.

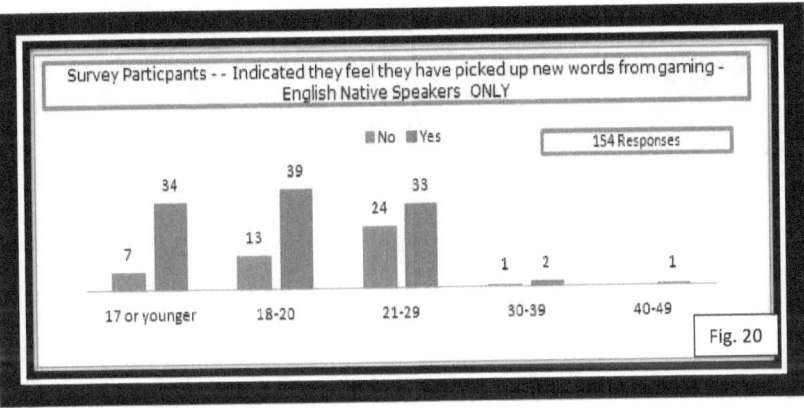

Native English speakers reported in the majority, that they felt they learned new words through video games.

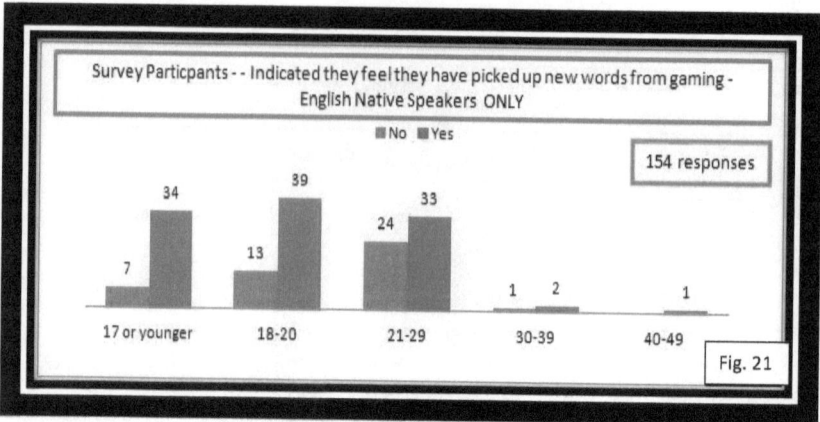

Fig. 21

The breakdown of the age brackets of the native speakers indicates that the 18 – 20 years old range responded in the majority; although, it should be noted that the levels are relatively equal across the range of yes's from the 17 – 29 age range.

Correlations

Hypothesis 1

There is no difference between time spent gaming and the participants Cambridge Score/self reported new words. (H_o)

With more time spent gaming the Cambridge Score will be higher/self reported new works score will be higher. (H_1)

Hours per Day Vs Learned New Words

			Learned New Words		Total
			Yes	No	
Hours per Day	0-4 hours per day	Count	35	5	40
		% within Hours per Day	87.5%	12.5%	100.0%
	2-4 hours per day	Count	218	15	233
		% within Hours per Day	93.6%	6.4%	100.0%
	3-4 hours per day	Count	132	5	137
		% within Hours per Day	96.4%	3.6%	100.0%
	8+ hours per day	Count	37	1	38
		% within Hours per Day	97.4%	2.6%	100.0%
Total		Count	422	26	448
		% within Hours per Day	94.2%	5.8%	100.0%

Symmetric Measures

		Value	Approx. Sig.
Nominal by Nominal	Cramer's V	.109	.150
N of Valid Cases		448	

While interpreting Cramer's V (Phi is valid only for tables 2 per 2) can be seen, p value is higher than 0.05 and the result is not statistically significant therefore the null hypotheses cannot be rejected.

Cambridge Vs. Hours per Day

Count

		Hours per Day				Total
		0-4 hours per day	2-4 hours per day	3-4 hours per day	8+ hours per day	
Cambridge	6-10	1	4	2	1	8
	11-12	0	6	2	0	8
	13-15	4	13	9	0	26
	16-17	4	23	17	3	47
	18-19	6	23	14	2	45
	20-21	4	50	27	3	84
	22-22	7	33	19	4	63
	23-25	15	86	48	25	174
Total		41	238	138	38	455

Symmetric Measures

		Value	Asymp. Std. Error[a]	Approx. T[b]	Approx. Sig.
Ordinal by Ordinal	Gamma	.088	.058	1.497	.134
N of Valid Cases		455			

While interpreting Gamma, the value is higher than 0.00 the correlation cannot be considered to be statistically significant and the null hypothesis cannot be rejected.

Hypothesis 2

There is no difference between the length of time that participants began gaming and participants Cambridge Score/self reported new Words. (H_0)

Length of time that Participants began gaming at would increase participants Cambridge score/ self reported new words. (H_1)

Cambridge Vs Age Started

Count

		Age Started				Total
		0-10 years old	11-20 years old	21-30 years old	31+ years old	
Cambridge	6-10	4	3	1	0	8
	11-12	6	1	1	0	8
	13-15	23	3	0	0	26
	16-17	35	11	1	0	47
	18-19	35	10	0	0	45
	20-21	62	22	0	0	84
	22-22	45	17	0	1	63
	23-25	141	33	0	0	174
Total		351	100	3	1	455

Symmetric Measures

	Value	Asymp. Std. Error[a]	Approx. T[b]	Approx. Sig.
Ordinal by Ordinal Gamma	-.086	.076	-1.106	.269
N of Valid Cases	455			

a. Not assuming the null hypothesis.
b. Using the asymptotic standard error assuming the null hypothesis.

As it can be seen in the above table, a very weak negative correlation (-0.086) is not statistically significant (p=0.269), therefore the null hypothesis stating that there is no connection between the age when the respondents started gaming and their English language knowledge cannot be refuted.

Age Started Vs Learned New Words

			Learned New Words		Total
			Yes	No	
Age Started	0-10 years old	Count	328	19	347
		% within Age Started	94.5%	5.5%	100.0%
	11-20 years old	Count	91	7	98
		% within Age Started	92.9%	7.1%	100.0%
	21-30 years old	Count	2	0	2
		% within Age Started	100.0%	0.0%	100.0%
	31+ years old	Count	1	0	1
		% within Age Started	100.0%	0.0%	100.0%
Total		Count	422	26	448
		% within Age Started	94.2%	5.8%	100.0%

Symmetric Measures

		Value	Approx. Sig.
Nominal by Nominal	Cramer's V	.036	.902
N of Valid Cases		448	

While interpreting Cramer's V it can be seen, the p-value is higher than 0.05 the correlation cannot be considered to be statistically significant and the null hypothesis cannot be rejected.

Hypothesis 3

Gaming Platform does not affect language learning. (H₀)

Gaming Platform effects language learning. (H₁)

Platform Used vs Learned New Words

			Learned New Words		Total
			Yes	No	
Platform Used	PC	Count	340	21	361
		% within Platform Used	94.2%	5.8%	100.0%
	Playstation	Count	54	1	55
		% within Platform Used	98.2%	1.8%	100.0%
	Wii/U	Count	9	1	10
		% within Platform Used	90.0%	10.0%	100.0%
	XBox	Count	19	3	22
		% within Platform Used	86.4%	13.6%	100.0%
Total		Count	422	26	448
		% within Platform Used	94.2%	5.8%	100.0%

Symmetric Measures

		Value	Approx. Sig.
Nominal by Nominal	Cramer's V	.099	.222
N of Valid Cases		448	

While interpreting Cramer's V it can be seen, the p-value is higher than 0.05 the correlation cannot be considered to be statistically significant and the null hypothesis cannot be rejected.

Cambridge Vs. Platform Used

			Platform Used				Total
			PC	Playstation	Wii/U	Xbox	
Cambridge	6-10	Count	3	3	1	1	8
		% within Cambridge	37.5%	37.5%	12.5%	12.5%	100.0%
	11-12	Count	6	2	0	0	8
		% within Cambridge	75.0%	25.0%	0.0%	0.0%	100.0%
	13-15	Count	13	9	0	4	26
		% within Cambridge	50.0%	34.6%	0.0%	15.4%	100.0%
	16-17	Count	29	10	0	8	47
		% within Cambridge	61.7%	21.3%	0.0%	17.0%	100.0%
	18-19	Count	31	9	3	2	45
		% within Cambridge	68.9%	20.0%	6.7%	4.4%	100.0%
	20-21	Count	67	10	2	5	84
		% within Cambridge	79.8%	11.9%	2.4%	6.0%	100.0%
	22-22	Count	55	6	2	0	63
		% within Cambridge	87.3%	9.5%	3.2%	0.0%	100.0%
	23-25	Count	163	7	2	2	174
		% within Cambridge	93.7%	4.0%	1.1%	1.1%	100.0%
Total		Count	367	56	10	22	455
		% within Cambridge	80.7%	12.3%	2.2%	4.8%	100.0%

Symmetric Measures

		Value	Approx. Sig.
Nominal by Nominal	Phi	.429	.000
	Cramer's V	.248	.000
N of Valid Cases		455	

While interpreting Cramer's V it can be seen, p value is higher than 0.05 and the result is not statistically significant therefore the null hypotheses cannot be rejected.

Hypothesis 4

Type of game does not affect language learning. (H₀)

Cambridge vs Games

			Games					Total
			MOBA	RPG	FPS	MMO	Action Adventure	
Cambridge	6-10	Count	1	2	3	0	2	8
		% within Cambridge	12.5%	25.0%	37.5%	0.0%	25.0%	100.0%
	11-12	Count	0	5	3	0	0	8
		% within Cambridge	0.0%	62.5%	37.5%	0.0%	0.0%	100.0%
	13-15	Count	4	14	2	1	5	26
		% within Cambridge	15.4%	53.8%	7.7%	3.8%	19.2%	100.0%
	16-17	Count	6	17	10	6	8	47
		% within Cambridge	12.8%	36.2%	21.3%	12.8%	17.0%	100.0%
	18-19	Count	6	13	12	1	13	45
		% within Cambridge	13.3%	28.9%	26.7%	2.2%	28.9%	100.0%
	20-21	Count	24	18	27	8	7	84
		% within Cambridge	28.6%	21.4%	32.1%	9.5%	8.3%	100.0%
	22-22	Count	17	11	24	5	6	63
		% within Cambridge	27.0%	17.5%	38.1%	7.9%	9.5%	100.0%
	23-25	Count	40	29	75	14	16	174
		% within Cambridge	23.0%	16.7%	43.1%	8.0%	9.2%	100.0%

Type of game effects language learning. (H$_1$)

While interpreting Cramer's V it can be seen, the p-value is higher than 0.05 the correlation cannot be considered to be statistically significant and the null hypothesis cannot be rejected.

Games Vs Learned New Words

Count

		Learned New Words		Total
		Yes	No	
Games	MOBA - e.g. League of Legends etc	93	4	97
	RPG - e.g. Dragon Age, Mass Effect etc..	104	3	107
	FPS - e.g. Counter Strike: GO, Call of Duty etc.	142	12	154
	MMO - e.g. World of Warcraft etc.	32	2	34
	Action Adventure - e.g. Mario, Uncharted	51	5	56
Total		422	26	448

Symmetric Measures

		Value	Asymp. Std. Error^a	Approx. T^b	Approx. Sig.
Nominal by Nominal	Cramer's V	.099			.357
N of Valid Cases		448			

While interpreting Cramer's V it can be seen, p value is higher than 0.05 and the result is not statistically significant therefore the null hypotheses cannot be rejected.

Hypothesis 5

Playing with others does not affect language learning. (H_0)

Playing with others effects language learning. (H_1)

Played With vs Learned New Words

			Learned New Words		Total
			Yes	No	
Played With	Single	Count	92	3	95
		% within Played With	96.8%	3.2%	100.0%
	Friends	Count	187	11	198
		% within Played With	94.4%	5.6%	100.0%
	Others On Line	Count	143	12	155
		% within Played With	92.3%	7.7%	100.0%
Total		Count	422	26	448
		% within Played With	94.2%	5.8%	100.0%

Symmetric Measures

		Value	Approx. Sig.
Nominal by Nominal	Cramer's V	.072	.316
N of Valid Cases		448	

While interpreting Cramer's V it can be seen, p value is higher than 0.05 and the result is not statistically significant therefore the null hypotheses cannot be rejected.

Cambridge Vs Played With

			Played With			Total
			Single	Friends	Others On Line	
Cambridge	6-10	Count	3	1	4	8
		% within Cambridge	37.5%	12.5%	50.0%	100.0%
	11-12	Count	3	4	1	8
		% within Cambridge	37.5%	50.0%	12.5%	100.0%
	13-15	Count	8	16	2	26
		% within Cambridge	30.8%	61.5%	7.7%	100.0%
	16-17	Count	16	13	18	47
		% within Cambridge	34.0%	27.7%	38.3%	100.0%
	18-19	Count	14	20	11	45
		% within Cambridge	31.1%	44.4%	24.4%	100.0%
	20-21	Count	16	39	29	84
		% within Cambridge	19.0%	46.4%	34.5%	100.0%
	22-22	Count	10	27	26	63
		% within Cambridge	15.9%	42.9%	41.3%	100.0%
	23-25	Count	28	81	65	174
		% within Cambridge	16.1%	46.6%	37.4%	100.0%
Total		Count	98	201	156	455
		% within Cambridge	21.5%	44.2%	34.3%	100.0%

Symmetric Measures

		Value	Approx. Sig.
Nominal by Nominal	Phi	.251	.012
	Cramer's V	.177	.012
N of Valid Cases		455	

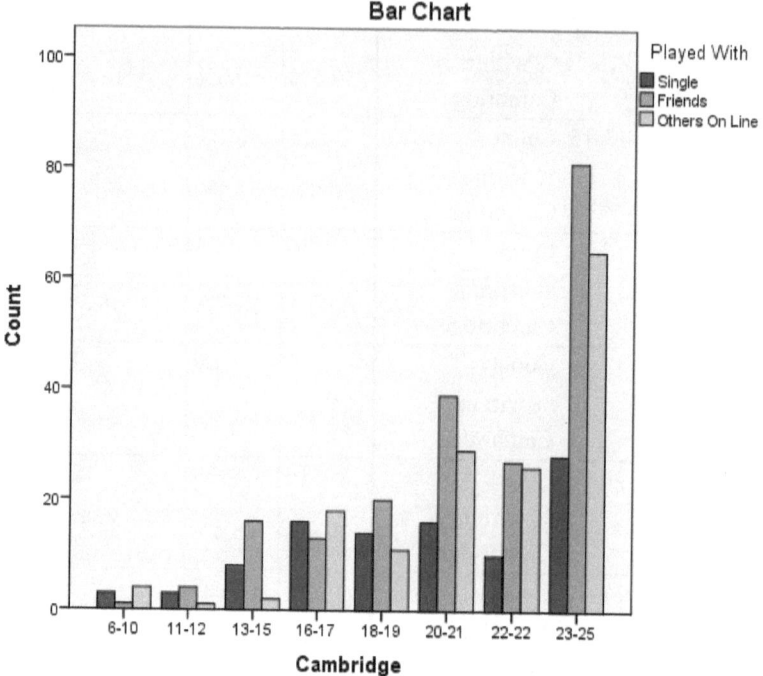

While looking at Crammer V we can notice a weak correlation (0.177) which is statistically significant (p = 0.012). On the basis of this result, we can say that there are statistically significant differences in English language knowledge between people who play with friends, alone, and others online.

Qualitative

In order to fully understand the complexities of the topic of language learning and video games, it is important to analysis the opinions of the people who are playing video games and who are involved in the gaming industry. For the purposes of this section I will be discussing the input given to me by professional streamers, general streamers and general video game players.

This section will be broken down into 3 different parts; a section discussing the opinions of general video game players, a section discussing the opinions of general streamers and a section for professional streamers. The rationale behind this is for the purpose of getting a broader view of video gaming from several different perspectives of gaming. There were six participants of the general video game players, two general streamers and two professional streamers. The criteria I used to differentiate the general streamers from the professional streamers is those who have been partnered with Twitch and those who have not.

Twitch Partners are described by Twitch as "Twitch Partners are an exclusive group of the world's most popular video game broadcasters, personalities, leagues, teams and tournaments." Being a twitch partner means that you earn a share of the revenue generated from the online gaming broadcasts that are created.

All of the interviews are through ethnomethodology/conversational analysis, the aim of this approach is to identify and explore the everyday processes of language learning and video gaming during conversations.

General Video Game Players

Each of the participants in the General Video Game Players section have been separated into Alpha, Beta, Delta, Gamma, Epsilon and Zeta in order to protect their anonymity. They were asked the same questions, though each of the responses were unique in their own way. (Appendix 2)

The participants reported that family was a major influence for their exposure to video games. The majority of the participants reported that they began playing video games with either their immediate or close family between the ages of four to eight years old. This would indicate that there was a strong positive reinforcement from the participants' families to partake in this activity.

There are strong differences in each of the participants' favourite/most memorable game growing up. Alpha; preferred the music and the pace of Sonic the Hedgehog. Beta; preferred games like Aladdin for the Megadrive because of its story telling. Gamma preferred Deux Ex because of its innovative mechanics and for its story lines. Delta preferred games like Counter Strike because of its interactive aspects. Epsilon preferred Space Quest for its type commands which for Epsilon began his foundation in language learning. Zeta preferred games like Legends of Zelda for its story telling.

An important thing to note is the responses when asked if their video game experiences growing up was in English all of the participants reported that all of their video games were in English, this would prove to be a more challenging prospect for games like Legend of Zelda or Deux Ex, as the primary focus of those games would be towards story telling rather than pure mechanics of the game. Alpha identified this saying that "so in the eighties all Video games were in English, of course at the time, most video games were not more complex than Super Mario, so you needed only to know the difference between Start Game and Game Over." What these responses would illustrate is that everyone has different reasoning behind playing video games and why they like certain games more than others.

When posed the question of whether or not they felt that being exposed to gaming from such a young age helped you in an academic environment, the general consensus felt that they had become more adjusted to the concept of English rather than going into it without any experience of words being spoken. Alpha said that because of gaming the experiences he had with video games helped with subjects in school. For example, Civilisation (a strategy war game) helped with learning history. Beta, mentioned feeling happy seeing words that had previously been seen in games coming up on tests and feeling more comfortable being around the words. Delta considered that the reason why he was able to skip a grade in his early school life was attributed to him being exposed to English at such a young age. So he considered himself to have a head start. Epsilon stated that while he does believe he picked up vocabulary and expressions from his exposure to video games from a young age he believed that it didn't have much of an impact in his academic life. Gamma believed that it was a combination of factors that helped improve his level of English, and he believed that gaming was a factor in this. Zeta stated that it was an undeniable factor in her learning of the English language.

When asked if the participants went to play games with the state of mind of learning the language not just for the purpose of playing the game, the responses were as follows. The majority of the participants agreed that they played the game for the game itself not for learning a language, Alpha and Gamma both stated that there was no other way to play the video games except to learn the language as they were not available in any other language other than English at the time. Alpha goes on to mention instances in his childhood where he would be sitting with a dictionary trying to translate the words he was seeing in order to progress in the game. Zeta, responded that it was a combination of both gaming and language that was the motivation for her, due to a ban on English in the family home it gave her an opportunity to learn English through this medium.

Alpha believed that it was an unconscious process; He stated that "if we come back to Monkey Island, I don't know if you're aware of this game. The game screen is split in two, you have the "game" and under it you have a list of word to interact with the game (and your inventory) I needed to know the difference between push and pull in order to complete the game thus, learning English words without knowing it." Beta believed something similar and the opinion that was given was "yet I think a picture is taken by our brain when a thing like that happens. And that even in another language, you understand what's happening and eventually you put words on it." Delta in comparison isn't so sure that playing video games was the primary factor in the learning of English. "I don't know if it necessarily was because of video games. Throughout school I've always done fairly well when studying languages such as German, and I've never played German video game or seen German movies as I had done with English....I'm not sure. I guess I picked up a few words but I don't know if that made a very big difference." Epsilon stated that it was down to repetition. "I think it was mostly repetition! And sometimes because it sounded cool...Like for instance, the text adventures forced me to repeat "open door", "search for key card", "retrieve cartridge from bookshelf" so many times that even after so many years I remember that! And then modern games such as Call of Duty well, they have people talking so I pick up the expression maybe because I feel they sound good I guess...and also repetition." As for what he believes happens that causes the learning, he stated he doesn't believe people are trying to work it out for himself. "When you're a kid you just pick it up but it like learning your own language...you're not aware of it, and at the same time they teach you those rules of grammar you're not able to state but you kinda know somehow." Gamma and Zeta stated that outright he believed that it was an unconscious process, Gamma saying that "it was not a conscious or directed effort."

When asked the question about which would be more beneficial in learning a language single player or multiplayer, each of the participants had different opinions. Alpha believed that single play experience was a passive one, in which you weren't actively engaged, while in multi player you are forced to communicate. Beta, thinks that multiplayer is more immersive, given that you have to speak to other players. "SP (Single Player) like MP (Multiplayer) game can give a lot, limitation are just in context or background." Delta believes that Multiplayer would be better because of exposure to the English language. Epsilon stated that "a multiplayer interactive game (?)

definitely. There's only so much you can learn from a fixed narrative however rich, whereas if you play online you get exposed to different accents, slang, colloquialisms." Gamma opinion on this topic was the opposite, that a single player experience would be far more engrossing. "I certainly think story rich single play experience are much more useful because they get you engrossed in the story or the world building or the exposition to the point where you're actively trying to understand the language, to learn more about the thing that fascinates you, without any external dictation." Zeta agrees that single player experiences have been the most beneficial to her but that may be because of her lack of exposure to multiplayer games.

General Streamers

Each of the participants in the General Video Streamers section have been separated into A and B in order to protect their anonymity. They were asked the same questions, though each of the responses was unique in their own way. (Appendix 3)

Both of the participants A and B are non native English speakers. A's native language is Dutch and B's native language is Vietnamese. Both participants have been streaming casually for 2 years. These streamers have the same philosophy in relation to their streams mainly for the purpose of having fun while focusing on viewer interaction. They also believe that it is possible to learn a language though video games, and they report that the most common language they come across in their daily lives is English. However, this is where the similarities between the two participants stop.

When asking A where he learned English, he stated that it was from the internet. "I started reading from a fan fiction site when I was 10-11 and that was fully English, so that started me down the path, then when I started playing WoW and other games, so you kind of learn to adapt I would say, and my friend and I would hop in on a skype call so that forces you to constantly speak English every day, which helps a lot!" While B stated that he began learning English in Elementary school.

While both streamers believe that it is possible to learn a language through video games both have different opinions about it. A believes that "there shouldn't be video games about language learning games, don't force it on people, they will learn it eventually, in my opinion." B talks about his own experiences in terms of learning through video games "I improved my speaking and communication skill just by streaming and talking about the games. Try to explain my thoughts and analysis in situations I faced in games has strengthened my English a lot."

When asking the participants how they felt the language learning happened their responses were as follows. A believed that " I'm assuming the like the video game, so one could assume that they'll play it several times, if you keep playing something, eventually you will understand what is being said and memorise it. Or they'll research to understand what's being said and remember it. That's at least what I did, when I found a word I didn't understand back then." B thought that "you should talk more and explain your thinking out loud. It does not need to be correct grammar, though. Also, listening and reading the in-game context contributes a lot to language learning"

On the topic of language learning and the impact it has on them as streamers, A reported that he has yet to come across anyone in his stream who does not have a basic grasp of English. B reports that "the more you interact the better you learn. At first, it might be difficult to express when you technically talk to yourself in front of a computer. But eventually, you will feel more comfortable doing so because you become more confident of the ability to speak with audiences."

When asked about the prospect of there being a relationship between streamers playing video games and interacting with their community and language learning the participants had these responses. A and B both believe that it would be the same as playing a video game themselves, and attributes watching a streamer causing the viewer to pick up words.

The participants believe that immersion is a critical point in language learning, both participants stating that there has to be an interest or a desire to play the game in order to become engrossed in it enough to learn.

When asked if they believed that someone plays a game in order to learn a language, both of the participants thought that the love of language would come from the enjoyment of the game and wanting to know more.

Professional Streamers

Each of the participants in the Professional Streamers section have been separated into their Streamer names. They were asked the same questions as the General Streamer group. (Appendix 3)

EdeMonster is a native English language Twitch streamer who describes himself as a variety caster. He has been streaming for 2 years. He describes his philosophy of his stream to "enjoy life and be as happy and positive as possible".

When asked if he thought it was possible to learn a language through video games he describes it as "completely possible," when asked about how he thought it could happen he stated that: "I've asked many of my viewers and they have all stated that they used video games as a medium to learn English or at least inspire them to learn English. How it happens more or less I believe comes from just the ability of someone wanting to learn and putting their minds to it."

In relation to the impact language learning has on him as a streamer he commented that "I play a lot of games that require reading of sorts whether it be reading the script for the characters that aren't voiced or some piece of information that involves the games somehow. I occasionally run into words I don't understand and love having the opportunity to learn them."

When asked if he believed there could be a relationship between streams playing video games, interacting with their viewers and language learning his opinion was "I don't think there would ever come a time where an English speaking caster could have that happen. I feel that caster that are bilingual could easily have an experience like this help their viewers that only speak foreign languages to learn English."

In terms of immersion, and its affect on language learning EdeMonster believes that everything has to come from an interest. "If someone doesn't want to take the extra mile to learn more then it's not possible for them to learn."

ManVsGame, an American veteran streamer has been streaming since 2010. ManVsGame describes his philosophy for his stream as to be #1 entertaining, but he describes it to be more like a "nightly talk show" a place where people could just hang out with him.

When discussing if it was possible to learn languages through mediums like video games, ManvsGame discusses his own experiences: "I learn a lot of Japanesse playing Sakura Wars, for example. I learned most of my Japanese from living in Japan for 10 months and being immersed in the language, but yes, I think video games can absolutely be a help tool in addition to more traditional forms of learning languages"

In relation to how he believes the learning happens he believes it comes down to the translation elements. "It's like translating anything else. The best kind of games for this one are the ones where you're able to pause and take your time to translate what is on screen and proceed through the game." He then provided me with an example of him doing just this in a Japanese game called Ni No Kuni, **http://www.gamefaqs.com/boards/998014-ni-no-kuni-wrath-of-the-white-witch/65208152**

When discussing whether or not he believes there is a relationship between streamers interaction with their viewers and language learning, he describes the process as being possible. "Anytime you are immersing yourself in a language, you're learning it...they're going to pick up some English."

Discussion

To begin with, I shall discuss the previous chapters and how they relate to the results of my work in order to answer the question: Is playing video games linked to level of English Language of the player? The introductory data established from my quantitative survey shows that there was a good spread of participants from around the world who took the survey. Fig. 7 illustrates the top 21 countries numerically. However it is important to note that participants from other countries outside the EU also responded. For example, the survey was responded to by participants from Taiwan, Korea, and Indonesia. This leads me to believe that there was a diverse group of people who participated in the survey who were non-native English speakers. This is important because the survey was to establish information about non-native English speakers.

What you will notice in Fig. 7 and Fig. 8 is that the major participating country was the United States of America and that there were native English speakers in this survey. This is, as far as I'm concerned, a result of over-enthusiasm on the participants' behalf. It seems likely that some respondents wished to be involved and, in their eagerness, they neglected to read the survey carefully enough. It would explain why in Fig. 2 you notice such a rapid drop off of numbers after question 7 of the survey. However, this does not mean that their overall submissions are irrelevant or invalid. They still provide very beneficial information in relation to language learning which should not be ignored.

Fig. 11 illustrates graphically the hours played per day versus the age range of the participants who completed the survey. It shows that, in all of the age groups, the majority of people play for about 2 – 4 hours a day and that as time goes on they will play less and less across all age groups. This is important when it comes to determining how long the participants actually spent engaged with the material and what impact this may have on their language ability as a result.

Fig. 12 shows the non-native English speakers' response to what language is most commonly spoken in games. The majority of participants responded that English was the predominant language in their daily gaming activity. This is important because in order to learn a language you need to be exposed to it. It could be argued that the participants of this survey should be exposed to 2 – 4 hours of English language a day. Fig. 13 shows that the majority of the participants have also been exposed to English Language either in school or in college. These are all good signs in relation to exposure to the language and the process of learning by immersion.

Fig. 15 shows that non-native English speakers report that they feel that they have learned new words through gaming. This is a useful self-reporting figure to consider, and it will be interesting to contrast this to both the Cambridge results, and what they self-report regarding their knowledge of English. Fig. 17 shows that most males report themselves to be in the Good category while females report themselves to be in the Excellent category. It should be pointed out at this stage that, due to the fewer number of female respondents, this self-reporting may not be representative of all female gamers. Fig. 18 shows the age range of the participants and reports what their language ability is.

The Cambridge scores, shown in Fig. 19, appear to support their claims. The majority of non-native English speakers in all categories scored in the 23 – 25 range of the Cambridge test. (Fig. 1). This means that the participants who scored in the range of 23 – 25 could potentially take a Cambridge proficiency test. As such, the participants' accuracy in self-reporting their language levels suggests that their self-reporting in terms of language learned from gaming, should be similarly accurate.

This is supported by Figs. 21 and 22 which show that the majority of non-native English speaking participants felt that they had learned new words from video gaming and the age range who agreed with this are the 17 – 39 age range. Additionally, this is also supported by Fig. 23 in which native English speakers report that they had felt that they had learned new words through video gaming.

How can these results be interpreted in relation to the theories that have been laid out in the introduction to this thesis? The results could potentially support the interaction and immersion theories. The data has stated that generally the majority of participants play for 2 – 4 hours a day. This exposure could lead to language learning by itself. The data also supports some of the case studies that have been discussed in this dissertation in which there have been reported cases of video gaming aiding language learning. The data itself, although self-reported, can add to the research already existing on this area. It will also lend to the merit of self-reporting as beneficial to the research into this area given the complexities in measuring this kind of research.

However this is in contrast to what is shown in the correlation data. So in order to understand why this has happened, a qualitative section was added to this thesis in order to understand the reasons for this.

The qualitative interviews offer three key elements in regards to the importance of video gaming and language learning from the participant's perspective. Immersion is one of the key elements that participants discussed in the interviews. Participants regarded immersion as being an important part of what stood out to them, what they liked to play and why they liked to play. All of the participants believed that being immersed in game play could be a potential factor for learning a language. This response links in with theoretical thinking already established in this dissertation.

Challenge is another key as to why the participants played video games. When asked why they played video games, all the participants said that they played because of the challenge that the game gave them. This element links with Gee (2005) in which he stated that people would not settle for games that were unchallenging. The element of challenge is also one of the elements that he believed enabled language learning.

Language is the primary key in relation to this dissertation and was the main topic of conversation for the participants. The way to discuss language in relation to video gaming is to look at the cultural element that kept appearing throughout the interviews. The participants were not all from the same country, therefore different customs came into play in relation to their language and how it reacts in their daily life. Family is another important aspect of this.

Family played a huge part in their participants' relationship with video gaming and language. Participants reported that their first instances with video gaming was with a family member or was inspired by a family member to participate in gaming. The influence of family connects to the young age at which the participants started gaming. Some of the participants detailed how siblings or other members of the family would sit with them and help them translate the game or explain meanings of the game.

This advancement of language that the participants reported to have gained from playing video games from such an early age, reportedly helped them in an academic environment. With some noted exceptions, some participants reported that while it helped them with some words and ideas, there was a big difference between the language that they could use in daily life vs. what they could say based on what they learned from a video game. This aligns with a point that Adams (2009) mentioned between the differences between applied literacy and literacy in games.

Participants unanimously said that in their youth the only gaming language that they were exposed to was English as there were no games available in their native languages at the time. This meant that if they wished to play the game, they had no choice but to play in English. This would show a love of two of the key elements that I have discussed previously; Challenge and Immersion.

All but one of the participants believed that language learning could happen through video gaming and that the process of learning happened unconsciously. The participants who thought that learning was possible thought that it was a by-product of playing the game because completing the game was the main objective of playing the game - not language learning. This would relate to the role of intent in the minds of the participants. It could be argued that, although they don't consciously realise it, there is an intent to learn the language. In order to play the game and become successful in completing the game they must learn the language. Therefore, intent is an active part of the learning process, albeit conscious or unconscious to the player.

The participant who said that they were uncertain about learning via video games reached this conclusion as a result of his/her interaction with other English language elements which were of influence in his/her youth, such as television, movies and music. He/she could not state with certainty that his/her early English language acquisition was solely down to video gaming.

I believe that this is a good point to have made, as it could not be completely ruled out that there may have been other sources that could have been influential to the participants at that time. However, for the purposes of this dissertation, it is important to try and focus down on a singular element, which is gaming. Whether or not it is the sole learning tool is not being discussed in this dissertation. What is being debated is whether there is a link between video gaming and the players' English ability. The data I have gathered shows that it is possible, and that there is a connection between video games and English language ability of the player.

When you look through all of the qualitative evidence you can see that there is a considerable gap between the opinion of professional players, amateur players and the general gaming public and what the quantitative data says about the process of language learning in video games. I do not believe that this is a gap that should be readily discarded.

I believe that there are other factors underlying this project that I have been unable to address that are causing such gaps within the data sets that I have gathered. I believe that the biggest misunderstanding is when it comes to the quantitative aspect is that the wrong measures are being used. While presently it is unclear what measures are incorrect we need to be sure that we are asking the right questions in order to get to the bottom of what is going on.

Understanding the prospective of both professional and amateur gamers alike opens up a whole universe worth of information in relation to what can be achieved. It is through this kind of research and understanding what the processes are; underlying thought and convention is the only way we are going to be able to progress to a level of truer understanding.

"However, in the end, the real importance of good computer and video games is that they allow people to re-create themselves in new worlds and achieve recreation and deep learning at one and the same time." Gee, 2003

Further Research

I believe that the data that has been collected here is only the tip of the iceberg in terms of the relationship between video gaming and English language learning. Additionally there are many other avenues that could be discussed such as the role of intent that has been touched upon in this dissertation. Cognitive approaches could be taken in relation to this as well. One suggestion for future research would be monitoring brain activation in people who are playing a video game, in their target or second language in comparison to those who are playing an educational language game to see if there is any difference in activation levels. A difficulty in this however would be in the testing of retention and its ecological value. Another aspect that could be considered in more detail is the effect of emotion on learning and gaming to show how the participants emotional state impacts and influences learning.

Bibliography

Adams, M. G. (2009). Engaging 21st-Century Adolescents: Video Games in the Reading Classroom. Retrieved June 13, 2013, from Literary Achievement Gap: http://literacyachievementgap.pbworks.com/f/EJ0986Engaging.pdf

Association, E. S. (2013). *Essential Facts About The Computer and Video Games Industry*. Retrieved March 14, 2013, from The Entertainment Software Association: http://www.theesa.com/facts/pdfs/ESA_EF_2013.pdf

Association, E. S. (2004). *Essential Facts About The Computer and Video Games Industry*. Retrieved March 14, 2004, from The Entertainment Software Association: http://www.theesa.com/facts/pdfs/ESA_EF_2004.pdf

Gee, J. P. (2005). *Good Video Games and Good Learning*. [online] Retrieved from: http://www.jamespaulgee.com/sites/default/files/pub/GoodVideoGamesLearning.pdf [Accessed: 25 Mar 2013].

Gee, J. (2003). What video games have to teach us about learning and literacy. *Computers In Entertainment (CIE)*, *1*(1), 20--20.

Green, P., Sha, M., & Liu, L. (2011). *The U.S.-China E-Language Project: A Study of a Gaming Approach to English Language Learning for Middle School Students*. Washington: RTI International/U.S Department of Education.

Merriam-webster.com. 2014. *Immersion - Definition and More from the Free Merriam-Webster Dictionary*. [online] Available at: http://www.merriam-webster.com/dictionary/immersion [Accessed: 20 Mar 2014].

Mitchell, R., Myles, F., & Marsden, E. (2013). *Second Language Learning Theories*. New York: Routledge.

Piirainen-Marsh, A., & Tainio, L. (2009). *Collaborative game-play as a site for participation and situation learning of a second language*. Retrieved March 9, 2013, from Taylor Francis Online: http://www.tandfonline.com/doi/abs/10.1080/00313830902757584#preview

Passer, M. W. 2009. *Psychology*. London: McGraw-Hill Higher Education.

Rankin, Y., Gold., R. & Gooch, B. (2006). *3D Role-Playing Games as Language Learning Tools*. [online] Retrieved from: http://webhome.cs.uvic.ca/~bgooch/Publications/PDFs/Rankin_Gold_Gooch.pdf [Accessed: 10 Mar 2014].

Reed, W. M., Kuwada, K., & deHaan, J. (2010, June). *The effect of interactivity with a musci video game on second language vocabulary recall*. Retrieved June 10, 2013, from Arizona Edu: www.u.arizona.edu/~ksaharty/Article D.html

Turgut, Y. & Irgin, P. (2009). *Young Learners' language learning via computer games*. [online] Retrieved from: http://www.gsedu.cn/tupianshangchuanmulu/zhongmeiwangluoyuyan/language%20learning%20via%20computer%20games.pdf [Accessed: 12 Mar 2014].

Appendix 1

Survey Questions:

1) Are you male or female?
2) Which category below includes your age?
3) Country of residence
4) What is your native language?
5) Have you ever had English lessons?
6) How would you rate your English ability?
7) Which of the following categories best describes your employment status?
8) What was your score on the test? (Please see above link and completely the short survey to get your score)
9) What age did you start playing video games at?
10) What platform do you primarily use for gaming?
11) What do you primarily play?
12) How long would you spend gaming per day
13) Who would you play video games with?
14) How open are you about your gaming life to friends and family?
15) What is the primary language used in the games you play?
16) Have you experienced English being spoken in your daily gaming activities?
17) If you have experienced English being spoken in your daily gaming activities, do you feel you have picked up some words from seeing/hearing them?

Appendix 2

Interview Questions for Survey participants

Note: Conversations happened via medium of Skype so transcriptions are taken directly from Skype.

1) I will ask you to start from the very beginning of your gaming life. Where did it all begin?
2) What was your favourite game growing up? What were the distinguishing features about the game?
3) So in later years, as you grew up, did your feelings for gaming ever change? Was it a constant in your life?
4) Did your tastes change or develop as you grew older - did you move away from [whatever game was previously discussed)?
5) What did you find so appealing about these games?
6) Were the games you were playing in your native language or were they in English?
7) When were you first exposed to English through video games?
8) Do you remember any experiences [in relation to emotion] from your gaming experiences?
9) Do you think that personal or emotional connection would be important in learning?
10) What were your feelings around learning a language in a schooling environment?
11) Do you think that you being exposed to gaming from such a young age helped you in an academic environment?
12) Why do you think you've picked up the language through playing video games?*
13) Did you go in with the intention of learning the language?
14) So would you say that you believe that it is an unconscious process and a by product of playing the game. Would that be a correct summarisation of what you're saying?*
15) What would you think are the limitations to using video gaming as an aid to learning a second language?
16) So what would you think about modern games, like Dark Souls, Fallout 3, Kingdom Hearts, final fantasy, that could have a very rich & detailed single player experience. Would you recommend these types of

games for people who lets say - didn't have an internet connection but who wanted to immerse themselves in a game but could only get a hold of an English version of a game?

17) What would be your opinion on games like these as a learning aid where multiplayer interaction isn't a necessity in order to complete the game or engage in it to the fullest?

Appendix 3

Interview Questions for Streamer participants

Note: Conversations happened via medium of Skype so transcriptions are taken directly from Skype. * indicates questions that were asked if applicable

1) What is the game you mainly stream and what is your native language?
2) How long have you been streaming for?
3) So where did you learn English?*
4) So how would you describe your twitch stream, what is your philosophy around it? What I mean by that is, some channels have an aim they want to achieve - for example ManvsGame's ethic is around beating a game and not stopping till he completes it, no spoilers etc.
5) Do you believe in your own experience that it is possible to learn a language through a medium like video games?
6) If you think people can learn a language through video games how do you think it happens?
7) So as a streamer, you interact with a lot of people on a daily basis, how do you feel as being a streamer language learning affects or impacts you as a streamer?
8) Do you feel as a streamer that it has helped you learn English better?*
9) Do you find that English seems to be the most common language spoken?
10) Twitch is an environment in which streamers all over the world can stream, but more often than not viewers depends on what time zone you're in (you can disagree with this if you feel I'm wrong in this assessment) so why do you think English is most widely spoken? Do you think it is in relation to the games that are being played?
11) Do you think that there could be a relationship between streamers playing video games, interacting with their viewers and language learning?
12) Do you believe that is down to immersion (ie. becoming really engrossed in the game) and if so, do you think that this applies to watching a stream? Do you think that this immersion is an important part of language learning?
13) Do you believe that if someone is playing a video game and learns English for example that they are playing a game for the purpose of learning the language?

I want morebooks!

Buy your books fast and straightforward online - at one of the world's fastest growing online book stores! Environmentally sound due to Print-on-Demand technologies.

Buy your books online at
www.get-morebooks.com

Kaufen Sie Ihre Bücher schnell und unkompliziert online – auf einer der am schnellsten wachsenden Buchhandelsplattformen weltweit! Dank Print-On-Demand umwelt- und ressourcenschonend produziert.

Bücher schneller online kaufen
www.morebooks.de

OmniScriptum Marketing DEU GmbH
Heinrich-Böcking-Str. 6-8
D - 66121 Saarbrücken
Telefax: +49 681 93 81 567-9

info@omniscriptum.com
www.omniscriptum.com

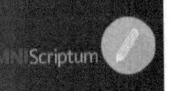

www.ingramcontent.com/pod-product-compliance
Lightning Source LLC
Chambersburg PA
CBHW031538210526
45464CB00003B/1058